Cambridge Elements ≡

Elements in Geochemical Tracers in Earth System Science
edited by
Timothy Lyons
University of California
Alexandra Turchyn
University of Cambridge
Chris Reinhard
Georgia Institute of Technology

THE TEX$_{86}$ PALEOTEMPERATURE PROXY

Gordon N. Inglis
University of Southampton

Jessica E. Tierney
University of Arizona

CAMBRIDGE
UNIVERSITY PRESS

CAMBRIDGE
UNIVERSITY PRESS

University Printing House, Cambridge CB2 8BS, United Kingdom

One Liberty Plaza, 20th Floor, New York, NY 10006, USA

477 Williamstown Road, Port Melbourne, VIC 3207, Australia

314–321, 3rd Floor, Plot 3, Splendor Forum, Jasola District Centre,
New Delhi – 110025, India

79 Anson Road, #06–04/06, Singapore 079906

Cambridge University Press is part of the University of Cambridge.

It furthers the University's mission by disseminating knowledge in the pursuit of
education, learning, and research at the highest international levels of excellence.

www.cambridge.org
Information on this title: www.cambridge.org/9781108810531
DOI: 10.1017/9781108846998

© Gordon N. Inglis and Jessica E. Tierney 2020

First published 2020

A catalogue record for this publication is available from the British Library.

ISBN 978-1-108-81053-1 Paperback
ISSN 2515-7027 (online)
ISSN 2515-6454 (print)

Cambridge University Press has no responsibility for the persistence or accuracy of
URLs for external or third-party internet websites referred to in this publication
and does not guarantee that any content on such websites is, or will remain,
accurate or appropriate.

The TEX$_{86}$ Paleotemperature Proxy

Elements in Geochemical Tracers in Earth System Science

DOI:10.1017/9781108846998
First published online: September 2020

Gordon N. Inglis
University of Southampton

Jessica E. Tierney
University of Arizona

Author for correspondence: Gordon N. Inglis, gordon.inglis@soton.ac.uk

Abstract: The TEX$_{86}$ paleothermometer is based on the distribution of archaeal membrane lipids (glycerol dialkyl glycerol tetraethers [GDGTs]) in marine sediments. GDGTs are ubiquitous, abundant and relatively resistant to degradation; as such, the TEX$_{86}$ paleothermometer has been used to reconstruct sea surface temperature during the Cenozoic and early Mesozoic. We review the principles of the TEX$_{86}$ proxy and developments made over the past two decades. We also discuss its application as a paleotemperature proxy and explore existing challenges and limitations.

Keywords: Archaea, glycerol dialkyl glycerol tetraethers, sea surface temperature,

ISBNs: 9781108810531 (PB), 9781108846998 (OC)
ISSNs: 2515-7027 (online), 2515-6454 (print)

Contents

1 Introduction

Accurately quantifying sea surface temperature (SST) during Earth's history is critical for understanding a range of key climate processes, such as climate sensitivity and polar amplification. The distribution of molecular fossils ('bio-markers') (Brassell et al., 1986; Schouten et al., 2002) are widely used as paleothermometers, because they derive from a specific biological source organism and can be preserved in the sedimentary record for millions of years. An organic paleothermometer is based on the principle that microbes will adjust their membrane fluidity, and thus lipid profile, in response to temperature ('homeoviscous adaptation'). This is typically achieved by altering the number and/or position of double bonds, branches and/or rings of the cell membrane lipids.

The $U^{K'}_{37}$ index, based on the distribution of long-chain (C_{37}) ketones (alkenones) produced by haptophyte algae, is one of the most widely used organic paleothermometers (Brassell et al., 1986). However, $U^{K'}_{37}$ is unable to reconstruct SSTs beyond 30°C (Schouten et al., 2002; Wuchter et al., 2004), limiting its application during past warm climates. Alkenones are also relatively uncommon in sediments older than 50 million years – in fact, the more unsaturated compounds that form the base of the paleothermometer might be an evolutionary adaptation to Cenozoic cooling (Brassell, 2014).

The TetraEther indeX of 86 carbons (TEX$_{86}$) fills the gaps left by the alkenone paleothermometer by having a larger dynamic range and being present farther back in geological time. TEX$_{86}$ is based on the distribution of isoprenoidal glycerol dialkyl glycerol tetraethers (isoGDGTs) in marine (and lake) sediments (Schouten et al., 2002). Within marine environments, isoGDGTs are thought to be mainly derived from ammonia-oxidising Thaumarchaeota that inhabit both the surface ocean and mesopelagic zone (Church et al., 2010; Schouten et al., 2002; Villanueva et al., 2015; Wuchter et al., 2004). As these compounds are ubiquitous, abundant and relatively resist-ant to degradation, the TEX$_{86}$ paleothermometer can be applied throughout the Cenozoic and well into the Mesozoic, extending our understanding of Earth's climate into very warm ('greenhouse') worlds. In this Element, we review the systematics of the TEX$_{86}$ proxy, new developments over the past two decades and its application towards reconstructing paleotemperatures, including outstanding concerns and limitations.

2 History and Systematics

IsoGDGTs are archaeal membrane lipids that consist of two C_{40} biphytane chains connected by ether bonds to a terminal glycerol group with a varying number of rings; in marine and lacustrine environments, this is typically up to four cyclopentane moieties and one cyclohexane moiety (Figure 1a). In living organisms, isoGDGTs have a variety of polar head groups (intact polar lipid GDGTs; e.g. hexose, dihexose, phospho-hexose and hexose-phosphohexose head groups; Figure 1b) (Schouten et al., 2008). Although the transformation of intact polar lipid (IPL) GDGTs to core GDGTs is assumed to proceed rapidly following cell death, in some locations IPL GDGTs persist in the geological record for up to 2.5 million years (Lengger et al., 2014a). However, IPL GDGTs do not appear to significantly impact core lipid-derived TEX_{86} values (Lengger et al., 2014b).

GDGTs were originally discovered in the 1970s and 1980s within cultures of (hyper)thermophilic and thermoacidophilic Archaea (de Rosa et al., 1977). Subsequent culture experiments using *Sulfolobus solfataricus* (De Rosa et al., 1980) and *Thermoplasma acidophilum* (Uda et al., 2001) demonstrated that the number of cyclopentane moieties increased at higher temperatures, resulting in a more densely packed and stable membrane. Later studies demonstrated that GDGTs were also ubiquitous in the marine environment (Hoefs et al., 1997) and could be derived from non-thermophilic archaea (e.g. *Candidatus Cenarchaeum symbiosum*) (DeLong et al., 1998). Developments in high-performance liquid chromatography–mass spectrometry (HPLC-MS) later allowed for direct analysis of GDGT core lipids (see Section 3) and confirmed the widespread occurrence of isoGDGTs with up to four cyclopentane moieties in mesophilic settings, including marine environments (Hopmans et al., 2000). Sinninghe Damsté et al. (2002) also identified an additional isoGDGT which included four cyclopentane moieties and one cyclohexane moiety – crenarchaeol – as well as a later eluting isomer (Sinninghe Damsté et al., 2002). It was proposed that the latter was a regioisomer of crenarchaeol (Sinninghe Damsté et al., 2002). However, recent work suggests that this may instead represent an isomer in which the cyclopentane moiety adjacent to the cyclohexane moiety possesses the uncommon *cis* stereochemistry (Liu et al., 2018, Sinninghe Damsté et al., 2018). Both compounds had not been identified in other archaeal strains but were later determined to be specific to the phylum Thaumarchaeota (formerly Marine Group I Crenarchaeota) (De la Torre et al., 2008; Pitcher et al., 2010). Lincoln et al. (2014) suggested that Marine Group II (MG-II) Euryarchaeota can also synthesise crenarchaeol (Lincoln et al., 2014). However, the production of crenarchaeol by MG-II Euryarchaeota has been heavily debated (Lincoln et al., 2014, Schouten et al., 2014; Zeng et al., 2019). The unique structure of crenarchaeol may help to prevent packing of core GDGTs within Thaumarchaeotal lipids and may be an adaptation to colder, non-hyperthermophilic environments (Sinninghe Damsté et al., 2002).

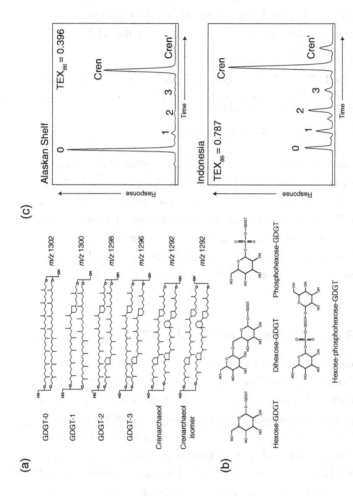

Figure 1 (a) Core structures of isoprenoidal GDGTs commonly found in marine and lake sediments, with the associated mass-to-charge ratio (*m/z*). (b) Common polar headgroups identified in IPL GDGTs which replace the terminal hydroxyl (–OH) group(s) shown in (a). (c) High-performance liquid chromatography–mass spectrometry (HPLC–MS) chromatograms of a 'cold' and 'warm' sedimentary GDGT distribution. Numbers correspond to the GDGT structures shown in (a). The upper panel shows a chromatogram from the Alaskan Shelf (Tierney, unpublished data). The bottom panel shows a chromatogram from the Makassar Strait, Indonesia (Tierney, unpublished data).

However, crenarchaeol has also been identified in cultures of thermophilic Thaumarchaeota (De la Torre et al., 2008) and terrestrial hot springs (Pearson et al., 2008); as such, its exact physiological role remains unclear.

Schouten et al. (2002) first suggested that the number of cyclopentane moieties in isoGDGTs within marine sediments was related to SST (Schouten et al., 2002). Sediments from colder regions contained mostly GDGT-0 and crenarchaeol, whereas warmer settings had a higher relative concentration of isoGDGTs 1 to 3 and the crenarchaeol isomer (Figure 1c). Through trial and error, this led to the formulation of the TEX_{86} index (Schouten et al., 2002):

$$TEX_{86} = GDGT\text{-}2 + GDGT\text{-}3 + Cren' / GDGT\text{-}1 + GDGT\text{-}2 + GDGT\text{-}3 + Cren'$$

This was related to SST using a linear calibration:

$$TEX_{86} = 0.015 \times SST + 0.28 \ (r^2 = 0.92, n = 44)$$

GDGT-0 and crenarchaeol were excluded from TEX_{86} because their inclusion dilutes the influence of other minor GDGTs (Schouten et al., 2002). In addition, GDGT-0 can also have additional sources (e.g. methanogenic archaea; see Section 6.1). The TEX_{86} paleothermometer was initially applied to mid-Cretaceous sediments (Schouten et al., 2003) and has since been used to reconstruct SST over the past 190 million years of Earth history (Cramwinckel et al., 2018; Inglis et al., 2015; O'Brien et al., 2017; Robinson et al., 2017).

3 Extraction and Analysis of GDGTs

Intact and core GDGTs are difficult to analyse via traditional gas chromatographic techniques because of their low volatility. As such, early studies utilised chemical degradation techniques to analyse their ether-bound hydrocarbon skeletons. This procedure involved refluxing the polar fraction with 56 wt % hydroiodic acid (HI) before reducing the alkyliodides to hydrocarbons. This enabled analysis of the ether-bound hydrocarbon skeletons via GC and GC-MS (DeLong et al., 1998; Hoefs et al., 1997). However, this process is laborious and it is not possible to identify the core structures of each GDGT from which they derive.

Instead, direct analysis of GDGT core lipids typically relies on HPLC/atmospheric pressure chemical ionisation–mass spectrometry (HPLC/APCI-MS). The HPLC-MS protocol was developed by Hopmans et al. (2000) and uses normal phase columns (e.g. amino, cyano or silica) and mixtures of hexane and isopropanol as the mobile phase (Hopmans et al., 2000). This method requires only simple extraction and purification methods, and round-robin studies indicate high reproducibility in TEX$_{86}$ values between laboratories (ca. 0.04 TEX$_{86}$ unit) (Schouten et al., 2013). Recent analytical advancements include (1) the use of single ion monitoring (SIM), which has enhanced the precision of TEX$_{86}$ measurements (Schouten et al., 2013) and (2) the use of ultra-high-performance liquid chromatography silica columns (Hopmans et al., 2016), which has enabled clearer chromatographic separation. For details on the current HPLC/APCI-MS operating conditions, see Hopmans et al. (2016). GDGT core lipids can also be analysed using high-temperature gas chromatography–mass spectrometry (Lengger et al., 2018). However, this method has yet to be adopted more widely.

4 Ecophysiology and Habitat of Thaumarchaeota

4.1 Ecophysiology

Early work noted that Thaumarchaeota represented 20–30% of all picoplankton cells in the world's ocean (Karner et al., 2001). However, insights into marine Thaumarchaeotal ecophysiology were initially limited. The stable carbon isotopic composition (δ^{13}C) of biphytanes (derived from GDGTs) from the marine water column and sediments indicated ^{13}C enrichment (ca. 3–4‰) relative to algal lipids derived from autotrophic organisms living close to the photic zone (e.g. cholestane; −24 to −27‰) (Hoefs et al., 1997), implying uptake of dissolved inorganic carbon (e.g. bicarbonate). This was consistent with biphytane radiocarbon values (Δ^{14}C), which were more similar to thermocline dissolved inorganic carbon (DIC; ^{14}C-depleted) than surface water DIC (^{14}C-enriched, due to the presence of 'bomb' ^{14}C) (Pearson et al., 2001). Experimental evidence for inorganic carbon uptake was confirmed by Wuchter et al. (2003), who documented significant uptake (ca. 70%) of ^{13}C-labeled bicarbonate into Thaumarchaeotal-derived biphytanes (Wuchter et al., 2003). The isolation and cultivation of a marine archaeon (*Nitrosopumilus maritimus* strain SCM1) definitively confirmed inorganic carbon uptake by marine Thaumarchaeota (Könneke et al., 2005). Qin et al. (2014) argued that cultured strains or natural Thaumarchaeotal communities could also incorporate organic substrates (e.g. short-chain acids). However, Kim et al. (2016) show that short-chain acids are not directly incorporated into ammonia-oxidising archaea (AOA) cells, but are instead involved in hydrogen peroxide detoxification.

The isolation and cultivation of *N. maritimus* also suggested conversion of ammonia to nitrite during Thaumarchaeotal cell growth (Könneke et al., 2005). This was consistent with earlier metagenomic studies which identified ammonium oxidation (AMO)-related genes in environmental sequences of marine Thaumarchaeota (Francis et al., 2005) and demonstrates that Thaumarchaeota are chemolithoautotrophic ammonium oxidisers. Recent work has also shown that Thaumarchaeotal strains (e.g. *N. maritimus*) can utilise cyanate and urea as alternative energy substrates through the intracellular conversion to ammonium (Kitzinger et al., 2019). As Thaumarchaeota are most abundant just below the upper photic zone (i.e. where nitrification rates are highest), they may play an important role in the nitrogen cycle.

4.2 Habitat and Depth of Export

Although the TEX_{86} proxy correlates strongly to SST or temperatures present between 0 and 200 m water depth, Thaumarchaeota can live throughout the water column (Karner et al., 2001). This implies that the export depth of TEX_{86} could extend below the photic zone. Analyses of intact polar GDGTs and Thaumarchaeotal genes (16S rRNA and *amoA*) suggest maximum abundance between 50 and 300 m, that is, within and just below the mixed layer of the ocean (Church et al., 2010; Schouten et al., 2012). A shallow subsurface origin is also consistent with $\delta^{13}C$ and $\Delta^{14}C$ analysis of core GDGTs and ether-cleaved biphytanes (Pearson et al., 2001; Shah et al., 2008). The meridional distribution of TEX_{86} data in the modern ocean also exhibits an inverted U shape, providing further evidence that TEX_{86} reflects subsurface (<200 m) ocean temperatures (Zhang and Liu, 2018). This work supports the general interpretation that TEX_{86} values reflect temperatures from the near surface (Yamamoto et al., 2012) or the shallow subsurface (dos Santos et al., 2010). Since there is a high correlation between subsurface temperatures and SSTs, TEX_{86} can generally still be used as an SST proxy even if GDGTs are produced lower in the mixed layer (Tierney and Tingley, 2014, 2015). However, in locations with a shallow thermocline, subsurface variations can be decoupled from SSTs, and in such situations TEX_{86} is better treated as a thermocline proxy (Wuchter et al., 2005).

It is assumed that biomass and GDGTs associated with archaea living within the deep subsurface (>1000 m) are not effectively exported to sediments due to the absence of effective export mechanisms (e.g. packaging onto faecal pellets or marine snow aggregates; e.g. Wuchter et al., 2005). However, this does not prevent a minor contribution of GDGTs from deeper Thaumarchaeotal sources. Indeed, several studies have observed changes in the distribution of isoGDGTs with increasing water depth (Taylor et al., 2013; Kim et al., 2015). This suggests the presence of a deep-water Thaumarchaeotal community with a fundamentally different isoGDGT distribution (Kim et al., 2015; Taylor et al., 2013).

Depth-related differences in 16S rRNA gene sequences (e.g. ammonia mono-oxygenase; *amoA*) support this observation (Villanueva et al., 2015) and indicate that marine Thaumarchaeota can be subdivided into two distinct clusters: (1) a shallow water cluster, abundant between <200 and approx. 500 m water depth and (2) a deep-water cluster, abundant in the deeper mesopelagic and bathypelagic ocean (typically >1000 m). Previous studies indicate that deeper waters are characterised by a higher abundance of GDGT-2 relative to GDGT-3 (and therefore a high GDGT-2/GDGT-3 ratio) (Taylor et al., 2013) and an increase in crenarchaeol isomer (Kim et al., 2015). If GDGTs from deeper waters are incorporated into the sedimentary GDGT pool, this can potentially lead to a warm bias in reconstructed TEX$_{86}$ SST estimates (up to 5°C) (Kim et al., 2015). However, practically speaking, this effect is rare and appears to significantly impact only SST estimates derived using TEX$_{86}{}^{L}$ (Inglis et al., 2015).

5 Preservation of GDGTs Within the Sedimentary Record

A key requirement for the application of TEX$_{86}$ in the sedimentary record is that GDGTs are resistant to degradation. Previous studies indicate that TEX$_{86}$ values are well preserved during grazing (Huguet et al., 2006) and vertical transport (Yamamoto et al., 2012). No significant effect was found under contrasting redox conditions, suggesting that oxic degradation also exerts a relatively minor influence on TEX$_{86}$ values (Schouten et al., 2004). However, a single study on the Madeira Abyssal Plain indicated a significant decrease in TEX$_{86}$-derived SSTs (up to 6°C) across an oxidation front (Huguet et al., 2009). This was attributed to selective preservation of terrestrial-derived isoGDGTs relative to marine-derived isoGDGTs (Huguet et al., 2009), but may also conceivably reflect a change in GDGT sources rather than differential degradation. In the Madeira Abyssal Plain, low TEX$_{86}$ values are associated with high BIT indices (due to selective degradation of isoGDGTs relative to brGDGTs), and a negative linear relationship between TEX$_{86}$ and BIT can potentially flag sediments impacted by oxic degradation. However, the impact of oxic degradation is not uniform and the direction of change in TEX$_{86}$ values will depend on the distribution of terrestrially derived isoGDGTs.

The TEX$_{86}$ paleothermometer can be influenced by thermal maturation. For example, hydrous pyrolysis experiments show a substantial decline in TEX$_{86}$ values above 240–260°C (Schouten et al., 2004). This is related to the preferential degradation of core GDGTs with an increasing number of cyclopentane moieties. This study confirms that the TEX$_{86}$ paleothermometer should not be applied in thermally mature settings (i.e. where hopane $\beta\beta/\alpha\beta + \beta\beta$ values <0.5

or hopane 22S/(22S + 22 R) ratios >0.1) and we recommend that authors report biomarker thermal maturity ratios alongside TEX_{86} values in older sediments (>50–100 Ma) so that any maturity effects can be assessed (O'Brien et al., 2017).

6 Constraints on the Application of TEX_{86}

6.1 Input from Exogenous Sources

Exogenous sources of GDGTs (i.e. those other than marine Thaumarchaeota) can complicate the interpretation of TEX_{86} values. However, several indices have been developed to screen for such secondary inputs. The branched-to-isoprenoidal tetraether (BIT) index (Hopmans et al., 2004) can be used to assess whether terrestrial soil input of isoGDGTs impacts TEX_{86}. Early work argued that TEX_{86} estimates with BIT values >0.4 should not be used for SST reconstruction (Weijers et al., 2006); however, the impact of terrestrial-derived isoGDGTs is highly variable and will depend on the nature and temperature of the source catchment (Inglis et al., 2015). Methanogenic archaea synthesise isoGDGTs with zero to three cyclopentane rings and thus can influence TEX_{86} values in places where there is substantial methane production (Blaga et al., 2009). The %GDGT-0 index is a qualitative indicator for the contribution of methanogenic archaea to the sedimentary GDGT pool: when %GDGT-0 values exceed 67%, an additional, potentially methanogenic, source of GDGT-0 is likely. Generally speaking, interference from methanogenic archaea is more of an issue for lacustrine settings (Blaga et al., 2009) than in marine sediments (Inglis et al., 2015). Methanotrophic Euryarchaeota can also synthesise GDGTs with zero to three cyclopentane moieties (Figure 2a) and can influence TEX_{86} values in settings dominated by anaerobic oxidation of methane (AOM; e.g. cold seeps, hydrothermal vents) (Zhang et al., 2011). The Methane Index (MI) can be used to assess the impact of anaerobic methanotrophy on TEX_{86} (Zhang et al., 2011), where low values (<0.3) indicate normal, marine conditions and high values (>0.5) indicate high rates of AOM. Consequently, TEX_{86} values should be excluded when MI values >0.5 (Figure 2a). However, high MI values do not necessarily indicate AOM (e.g. Polik et al., 2018) and should be accompanied by independent evidence for enhanced methane cycling (e.g. $\delta^{13}C_{GDGT}$).

The Ring Index (RI) – which represents the weighted average of cyclopentane moieties in GDGT compounds – can also distinguish soil and/or methane-impacted samples (Zhang et al., 2016). It can also help to quantify the extent to which samples deviate from the modern TEX_{86}–RI relationship (ΔRI). Zhang et al. (2016) argue that samples with ΔRI values >0.3 may indicate potentially

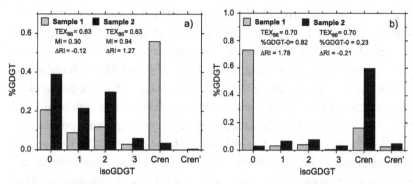

Figure 2 Interrogating GDGT distributions. (a) Two samples with identical TEX$_{86}$ values (= 0.63) but different Methane Index (MI) values. Sample 1 is a 'normal' marine sediment sample from the Arabian Sea (Turich et al., 2007). Sample 2 is a gas-hydrate impacted sediment sample from the Gulf of Mexico (Zhang et al., 2011). (b) Two samples with identical TEX$_{86}$ values (= 0.70) but different %GDGT-0 values. Sample 1 is a methanogen-impacted sediment sample from ODP Site 925 (Eocene-aged) (Inglis et al., 2015). Sample 2 is a 'normal' marine sediment sample from Mid-Waipara (Eocene-aged) (Hollis et al., 2012).

problematic TEX$_{86}$ values. To illustrate this, Figure 2b shows two sets of GDGT distributions with identical TEX$_{86}$ values (0.70) but different ΔRI values. Sample 1 has a high ΔRI value (>0.3), indicating an atypical GDGT distribution (Figure 2b). This sample is also characterised by high %GDGT-0 values (0.82), suggesting an additional methanogenic source of isoprenoidal GDGTs. This highlights the need to investigate the entire GDGT distribution before reconstructing SST.

6.2 Seasonality

TEX$_{86}$ values may be influenced by seasonal production (and export) of GDGTs; however, the evidence to date suggests that seasonality does not impact this proxy very strongly, perhaps because TEX$_{86}$ integrates the mixed layer (down to 200 or 300 m). Early sediment trap investigations suggested a cold (winter) bias in the North Sea (Herfort et al., 2006), consistent with the dominance of Thaumarchaeota in the winter season. In contrast, Wuchter et al. (2006) reported a warm (summer) bias in TEX$_{86}$ values within the Arabian Sea; however, a warm bias was restricted to the upper 500 m and seasonal variability was not present in the deeper sediment traps (Wuchter et al., 2006). Sediment trap studies from the Gulf of Mexico and Gulf of California also show little intra-annual variability, suggesting homogenisation of the seasonal cycle

(Richey and Tierney, 2016; Yamamoto et al., 2012). A warm (summer) bias has been reported from the Mediterranean (Huguet et al., 2011). However, sedimentary GDGTs in the Mediterranean basin may have a contribution from deep water Thaumarchaeota (Kim et al., 2015; Taylor et al., 2013), which can lead to a warm bias in reconstructed TEX$_{86}$ SST estimates (up to 5°C; see Section 4.2). In paleoclimate studies, a warm summer bias (especially in the high latitudes) is often cited to account for mismatches between other temperature proxies (e.g. δ^{18}O, Mg/Ca) or climate model simulations (Hollis et al., 2012). However, the aforementioned evidence indicates that seasonality has a relatively limited effect on sedimentary TEX$_{86}$ values. Indeed, the TEX$_{86}$–temperature relationship is not improved when using seasonal mean ocean temperatures (Kim et al., 2008) and calibration residuals are not suggestive of a strong seasonal influence (Tierney and Tingley, 2014).

6.3 Physiological Controls

TEX$_{86}$ can also be influenced by physiological and environmental controls. Batch culture experiments show an increase in TEX$_{86}$ values during late growth stage (Elling et al., 2014) and at lower O$_2$ concentrations (Qin et al., 2015). Both coincide with a decrease in ammonium oxidation rates, implying that variations in energy supply exert a control on TEX$_{86}$ values (Zhou et al., 2019). Radioisotope probing (Evans et al., 2018) and chemostat experiments (Hurley et al., 2016) also show that GDGT cyclisation is strongly influenced by NH$_4^+$ availability, with an increase in TEX$_{86}$ values under lower NH$_4^+$ concentrations. However, these findings are difficult to reconcile with experimental studies which demonstrate that ammonia oxidation rate is not sensitive to temperature between 8 and 20°C (Horak et al., 2013). Although such physiological factors remain difficult to assess in the geological record, the RI (Zhang et al., 2016) can possibly reveal non-thermal controls and thereby minimise such an influence.

7 Calibration of TEX$_{86}$ to SST

TEX$_{86}$ can be calibrated to SST using (1) an experimental approach based on cultures or mesocosm experiments and/or (2) an empirical approach based on core-top sediments. In Sections 7.1 and 7.2 we discuss the two main approaches.

7.1 Culture and Mesocosm Experiments

Early studies used mesocosm experiments to determine the relationship between TEX$_{86}$ and SST. Wuchter et al. (2004) incubated surface waters between 5 and 25°C and demonstrated that higher temperatures were associated with a linear increase in cyclisation (i.e. higher TEX$_{86}$ values) (Wuchter et al.,

2004). Schouten et al. (2007) also demonstrated a linear increase in cyclisation between 25 and 40°C (Schouten et al., 2007). Both mesocosm experiments had a similar linear slope and *y*-intercept, indicating a consistent relationship between TEX$_{86}$ and SST (Schouten et al., 2007; Wuchter et al., 2004). However, in both experiments, nutrient conditions were not controlled and therefore likely varied widely. In addition, the fractional abundance of the crenarchaeol isomer was approx. 14-fold lower than expected (Schouten et al., 2007; Wuchter et al., 2004). As a result, mesocosm-based TEX$_{86}$ calibrations have not been used for paleoclimate applications. The lack of crenarchaeol isomer is difficult to explain given its elevated abundance in warm, tropical settings (Kim et al., 2008) and in geological sediments (Inglis et al., 2015; O'Brien et al., 2017). Since *Nitrosopumilus maritimus* (belonging to Group I.1a Thaumarchaeota) does not synthesise the crenarchaeol isomer in high abundance (Elling et al., 2014), this may suggest that another species is responsible for its production in natural marine environments. Intriguingly, Group I.1b Thaumarchaeota can synthesise large quantities of crenarchaeol isomer (Pitcher et al., 2010); however, these organisms are usually not abundant within marine settings. Alternatively, crenarchaeol isomer could be derived from an additional deep-water source. This is supported by (1) Δ^{14}C analysis, which indicates an aged or non-surficial source for crenarchaeol isomer (Shah et al., 2008) and (2) an increase in the fractional abundance of crenarchaeol isomer within enclosed basins (e.g. Red Sea, Mediterranean) (Kim et al., 2015). However, additional work is required to test this hypothesis and will help to determine whether the isomer should continue to be included within TEX$_{86}$.

Batch cultures of marine Thaumarchaeota have also helped to isolate the influence of temperature on TEX$_{86}$ values. Elling et al. (2014, 2015) demonstrated that the cyclisation of GDGTs increased with growth temperature in three archaeal strains (*N. maritimus*, NAOA6 and NAOA2) (Elling et al., 2014, 2015). This was due primarily to changes in the relative abundance of GDGT-0 and Crenarchaeol. TEX$_{86}$ values increased linearly with temperature in two strains (*N. maritimus*, NAOA6) but not in the third (NAOA2), suggesting that changes in cyclisation may be species specific (Qin et al., 2015). As mesocosm- and culture-derived TEX$_{86}$ calibrations do not align with the core-top TEX$_{86}$ dataset, they have not been utilised in palaeoclimate studies.

7.2 Surface Sediment Calibrations

Multiple TEX$_{86}$ calibrations have been developed to estimate sea surface or shallow subsurface temperatures. The original TEX$_{86}$ core-top calibration was a linear relationship to SST (Schouten et al., 2002) (Figure 3). However, there is a weak relationship between TEX_{86} and SST <5°C, while sediments from the Red Sea yield much warmer TEX_{86}-derived SST estimates than observed values (Kim et al., 2008). As such, TEX_{86} was recalibrated to remove samples from the Red Sea and sites where SST <5°C (Kim et al., 2008) or <15°C (O'Brien et al., 2017). Non-linear calibrations have also been proposed (Kim et al., 2010; Liu et al., 2009). The most commonly applied is TEX_{86}^{H}, which uses the same combination of GDGTs as the original TEX_{86} but assumes an exponential relationship between temperature and SST (Kim et al., 2010) (Figure 3). This approach employs a log transformation and excludes data from the Red Sea and sites where SST <5°C. There is no experimental evidence to support (or refute) a log transformation. However, the 'reduced' dataset is not skewed (skewness = 0.007), suggesting there is no data-driven rationale for applying a log trans-formation (Hollis et al., 2019). Crucially, the TEX_{86}^{H} calibration does not account for observational error and is affected by regression dilution (Tierney and Tingley, 2014). This causes TEX_{86}^{H} to underestimate SSTs in the low latitudes (e.g. Cramwinckel et al., 2018; Inglis et al., 2015) as well as overesti-mate SSTs in the high latitudes. For this reason, we do not recommend using the TEX_{86}^{H} calibration (see Hollis et al., 2019). Kim et al. (2010) also proposed another index, TEX_{86}^{L}, which is also an exponential calibration but uses the entire core-top dataset (with the exception of the Red Sea) (Kim et al., 2010). TEX_{86}^{L} employs a combination of GDGTs that is different from TEX_{86}^{H}, removing GDGT-3 from the numerator and excluding crenarchaeol isomer entirely (Kim et al., 2010). However, this index lacks an obvious biological rationale and is particularly sensitive to subsurface export (Taylor et al., 2013). Its application within the geological record has been questioned (Inglis et al., 2015; Taylor et al., 2013) and thus we do not recommend its use.

The original TEX_{86} index has also been calibrated to SST using a Bayesian, spatially varying regression model (BAYSPAR) (Tierney and Tingley, 2014, 2015) (Figure 3). BAYSPAR assumes a linear relationship between TEX_{86} and SST but allows the regression parameters to vary spatially to accommodate observed changes in TEX_{86} sensitivity in locations such as the Mediterranean and the Red Sea. In the Late Quaternary, BAYSPAR uses the location of the site to determine which regression parameters to use. In deep-time settings, BAYSPAR searches the modern core-top dataset for TEX_{86} values that are similar to the measured TEX_{86} value within a user-specified tolerance and draws regression parameters from these modern 'analogue' locations. This approach also yields uncertainty bounds that reflect spatial differences in the slope and intercept terms and the error variance of the regression model and can be used to estimate uncertainty beyond the modern calibration range. Unlike previous models, BAYSPAR includes all modern TEX_{86} observations below

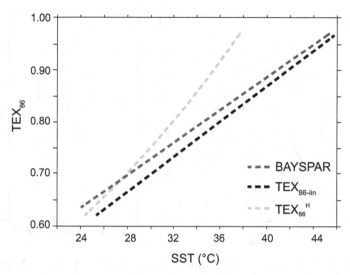

Figure 3 Comparison of the exponential TEX_{86}^{H} calibration (Kim et al., 2010), the linear TEX_{86} (O'Brien et al., 2017) and BAYSPAR calibrations (Tierney and Tingley, 2014, 2015). Owing to the spatially varying regression used for BAYSPAR, only one site (Eocene-aged ODP Site 959) is plotted here.

70°N, including the Red Sea. The inclusion of the Red Sea may be important given that Red Sea–type GDGT distributions are often observed in many geological settings (Hollis et al., 2019). Both surface ($BAYSPAR_{SST}$) and subsurface ($BAYSPAR_{SubT}$) calibrations are available, the latter being the gamma-weighted average of the temperature range over 0–200 m water depth, with a maximum probability at approx. 50 m (Tierney and Tingley, 2015).

More recently, Eley et al. (2019) proposed using Gaussian process regression (GPR) to calibrate GDGTs to SST (OPTiMAL; Eley et al., 2019). OPTiMAL has great potential for identifying anomalous GDGT distributions using a nearest neighbour distance metric (i.e. $D_{nearest}$); however, it is agnostic about the nature of GDGT response to temperature. As such, it is (currently) unable to extrapolate TEX_{86} beyond the modern calibration range.

8 Example Applications of TEX₈₆

8.1 Paleocene-Eocene Thermal Maximum

The TEX_{86} paleothermometer has been widely applied to the Paleogene (66 to 23 million years ago; see Section 8.2). However, many studies have focused on the Paleocene-Eocene Thermal Maximum (PETM; 56 Ma), a rapid global

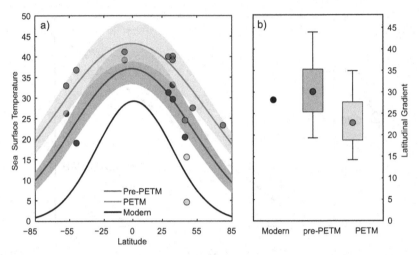

Figure 4 TEX$_{86}$-derived SST estimates for the pre-PETM and PETM time intervals. (a) pre-PETM (in blue) and PETM (in red) TEX$_{86}$ data and inferred latitudinal gradients from Gaussian fits to the data. Outlying data from the Fur section (North Sea) are plotted in grey and not included in the Gaussian fits. See Hollis et al., 2019 for full dataset. (b) Box plots of inferred pole-to-equator latitudinal temperature gradients for each time interval, derived from Gaussian fits to the posterior ensemble of calibrated SST values using BAYSPAR. Red dots denote median values, boxes bound the 25–75% ranges, and whiskers enclose the 90% confidence interval.

warming event associated with the release of ^{13}C-depleted carbon into the ocean–atmosphere system. Zachos et al. (2006) originally applied TEX$_{86}$ to a mid-latitude continental margin setting located in New Jersey, USA (35°N paleolatitude) (Zachos et al., 2006). TEX$_{86}$ indicated substantial warming during the PETM (9°C when recalibrated using BAYSPAR), consistent with planktic foraminifera δ^{18}O-derived SST estimates from the same site (ca. 6–9°C) (Zachos et al., 2006). TEX$_{86}$ has subsequently been applied to other mid-latitude (30–60°) and high-latitude (>60°) settings (Hollis et al., 2019 and references therein) and indicates a similar magnitude of warming during the PETM (ca. 8–11°C when recalibrated using BAYSPAR; Hollis et al., 2019) (Figure 4a).

TEX$_{86}$ has also been applied to several low-latitude PETM-aged sites (<30°) (Cramwinckel et al., 2018; Frieling et al., 2017). These studies illustrate that SSTs were >35°C during the PETM, refuting the existence of a strict tropical thermostat (Frieling et al., 2017). These studies also indicate a lower magnitude of warming (ca. 4–7°C when recalibrated using BAYSPAR) (Figure 4a), implying a weaker latitudinal temperature gradient (i.e. the temperature differ-ence between the low- and high-latitudes) during the PETM (ca. 10–15°C) (Figure 4b). Previous studies have shown that latitudinal temperature gradients of <20°C are difficult for climate models to simulate and require large changes in latitudinal heat transport and/or substantial positive feedbacks acting at high latitudes (Inglis et al., 2015). However, recent progress has been made via changes in the physical parameters of the model (e.g. cloud parameters) (Sagoo et al., 2013).

An important caveat is that many PETM sites are characterised by TEX_{86} values that exceed the calibration range in modern oceans (0.3–0.8). It is likely that these higher values indicate higher temperatures, but the degree of esti-mated warming depends on assumptions about the mathematical nature of the TEX_{86}-temperature relationship (e.g. linear, exponential; see Sections 7.1 and 7.2). Batch culture and chemostat experiments demonstrate that cyclisation of GDGTs responds in a linear fashion up to temperatures of 40°C (i.e. TEX_{86}-linear, BAYSPAR). However, culture experiments may not have the same (linear) relationship between different strains. That means that mixing of communities due to population changes could easily be non-linear. As such, a non-linear relationship remains possible.

8.2 Early Cenozoic Cooling

The generation of long-term, regional TEX_{86} records has been essential for developing a more detailed picture of global cooling during the early Cenozoic (56 to 34 million years ago) and elucidating the driving mechanisms responsible (i.e. CO_2 vs. ocean circulation). Pearson et al. (2007) originally applied TEX_{86} to sediments from Tanzania. This study indicated stable and warm SSTs during the early Cenozoic (Pearson et al., 2007). However, recent evidence from other low-latitude sites indicate substantial surface water cooling during this interval (Cramwinckel et al., 2018; Inglis et al., 2015) (ca. 8–10°C of cooling when recalibrated using BAYSPAR) (Figure 5a). TEX_{86}-derived SST estimates from the high-latitudes also suggest surface water cooling during the early Cenozoic (ca. 10–12°C of cooling when recalibrated using BAYSPAR) (Bijl et al., 2009; Hollis et al., 2012). Remarkably, both low-latitude and high-latitude TEX_{86}-derived SST estimates decline in tandem and closely mimic global climate trends (Figure 5b). This evidence, in combination with existing CO_2 estimates (Anagnostou et al., 2016) and modelling simulations (Inglis et al., 2015), strongly suggests that drawdown of CO_2, rather than changes in ocean circula-tion, was the primary forcing for long-term climatic cooling during the early Cenozoic.

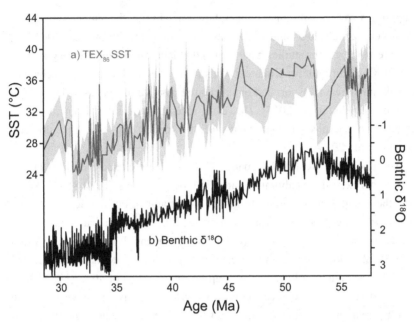

Figure 5 (a) TEX$_{86}$-derived, low-latitude SST estimates (ODP Site 929, ODP Site 925, ODP Site 959 and Tanzania) during the early Cenozoic (56 to 34 million years ago) (Cramwinckel et al., 2018). SSTs determined using BAYSPAR calibration (mean = 28, standard deviation = 10, search tolerance = 0.15). (b) Benthic foraminifera $\delta^{18}O$ values during the early Cenozoic (56 to 34 Ma).

9 Future Developments in TEX$_{86}$ Paleothermometry

Over the past 20 years, the TEX$_{86}$ paleothermometer has been applied to marine sediments spanning the last 190 million years of Earth's history. This has led to important inferences concerning past temperature variability and Earth system sensitivity (Cramwinckel et al., 2018; Inglis et al., 2015; O'Brien et al., 2017). However, several uncertainties still remain. The recent identification of two enzymes required for pentacyclic ring formation suggests that Thaumarchaeota are the dominant source of cyclised GDGTs in the open ocean (Zeng et al., 2019). However, the importance of (deep) subsurface GDGT export on the sedimentary GDGT pool remains unclear. The analyses of archaeal diversity (e. g. 16S rRNA gene analysis) (Villanueva et al., 2015) and intact polar lipid GDGT distributions in suspended particular matter (SPM) and surface sediments may help to improve our understanding of export dynamics in modern systems. The exact relationship between TEX$_{86}$ and temperature (especially beyond the modern calibration range) also remains debated. This requires an improved understanding of Thaumarchaeotal ecology. Continuous culture (chemostat) experiments are likely to be highly valuable and (unlike batch cultures) can isolate the influence of a single variable (e.g. temperature, oxygen, growth rate) upon the lipid composition (Hurley et al., 2016). Spooling Wire Microcombustion (SWiM)-isotope ratio mass spectrometry (IRMS) (Pearson et al., 2016) and/or high-temperature gas chromatography-isotope ratio mass spectrometry (HT/GC-IRMS) (Lengger et al., 2018) also provide the unique opportunity to analyse the carbon isotope analysis of core GDGTs and obtain unique new insights into archaeal ecology.

The development of new analytical techniques also has the potential to revolutionise the TEX$_{86}$ paleothermometer. The application of matrix-assisted laser desorption/ionisation coupled to Fourier transform-ion cyclotron resonance-mass spectrometry (MALDI-FT-ICR-MS) has recently enabled detection of GDGTs at ultra-high spatial resolution (<200 μm) (Wörmer et al., 2014). This method can provide high-resolution GDGT records within laminated or varved sediments. However, it is currently unable to differentiate between crenarchaeol and its isomer; as such, SST reconstruction using TEX$_{86}$ is currently not possible.

A major outstanding challenge is the large uncertainty in TEX$_{86}$-derived SSTs in past warm climates, as the functional behaviour of GDGT ratios at high temperature remains unknown. The development of new indices or calibrations could help to reduce the uncertainty associated with TEX$_{86}$-derived SST estimates (e.g. Bayesian regression models) (Tierney and Tingley, 2014). However, experimental approaches (e.g. batch or continuous culture experiments) are required to decipher the functional form of TEX$_{86}$ in the upper temperature range. There also remains significant value in employing TEX$_{86}$ alongside other SST proxies (e.g. δ^{18}O, Δ47, Mg/Ca), as this allows more accurate and robust quantification of temperature trends and anomalies.

References

Anagnostou, E., John, E. H., Edgar, K. M., et al. (2016) Changing atmospheric CO_2 concentration was the primary driver of early Cenozoic climate. *Nature* **533**, 380–4.

Bijl, P. K., Schouten, S., Sluijs, A., Reichart, G.-J., Zachos, J. C. and Brinkhuis, H. (2009) Early Palaeogene temperature evolution of the southwest Pacific Ocean. *Nature* **461**, 776–9.

Blaga, C. I., Reichart, G.-J., Heiri, O. and Sinninghe Damsté, J. S. (2009) Tetraether membrane lipid distributions in water-column particulate matter and sediments: A study of 47 European lakes along a north–south transect. *Journal of Paleolimnology* **41**, 523–40.

Brassell, S., Eglinton, G., Marlowe, I., Pflaumann, U. and Sarnthein, M. (1986) Molecular stratigraphy: A new tool for climatic assessment. *Nature* **320**, 129.

Brassell, S. C. (2014) Climatic influences on the Paleogene evolution of alkenones. *Paleoceanography* 29(3), 255–72. https://doi.org/10.1002 /2013PA002576

Church, M. J., Wai, B., Karl, D. M. and DeLong, E. F. (2010) Abundances of crenarchaeal amoA genes and transcripts in the Pacific Ocean. *Environmental Microbiology* **12**, 679–88.

Cramwinckel, M. J., Huber, M., Kocken, I. J., et al. (2018) Synchronous tropical and polar temperature evolution in the Eocene. *Nature* **559**, 382.

De la Torre, J. R., Walker, C. B., Ingalls, A. E., Könneke, M. and Stahl, D. A. (2008) Cultivation of a thermophilic ammonia oxidizing archaeon synthesizing crenarchaeol. *Environmental Microbiology* **10**, 810–18.

DeLong, E. F., King, L. L., Massana, R., et al. (1998) Dibiphytanyl ether lipids in nonthermophilic crenarchaeotes. *Applied and Environmental Microbiology* **64**, 1133–8.

De Rosa, M., de Rosa, S., Gambacorta, A., Minale, L. and Bu'lock, J. D. (1977) Chemical structure of the ether lipids of thermophilic acidophilic bacteria of the *Caldariella* group. *Phytochemistry* **16**, 1961–5.

De Rosa, M., Esposito, E., Gambacorta, A., Nicolaus, B. and Bu'Lock, J. D. (1980) Effects of temperature on ether lipid composition of *Caldariella acidophila*. *Phytochemistry* **19**, 827–31.

dos Santos, R. A. L., Prange, M., Castañeda, I. S., et al.(2010) Glacial–interglacial variability in Atlantic meridional overturning circulation and thermocline adjustments in the tropical North Atlantic. *Earth and Planetary Science Letters* **300**, 407–14.

Eley, Y.L., Thompson, W., Greene, S.E., Mandel, I., Edgar, K., Bendle, J.A. and Dunkley Jones, T., 2019. OPTiMAL: A new machine learning approach for GDGT-based palaeothermometry. *Climate of the Past Discussions*, pp.1–39. https://doi.org/10.5194/cp-2019-60

Elling, F. J., Könneke, M., Lipp, J. S., Becker, K. W., Gagen, E. J. and Hinrichs, K.-U. (2014) Effects of growth phase on the membrane lipid composition of the thaumarchaeon *Nitrosopumilus maritimus* and their implications for archaeal lipid distributions in the marine environment. *Geochimica et Cosmochimica Acta* **141**, 579–97.

Elling, F. J., Könneke, M., Mußmann, M., Greve, A., and Hinrichs, K.-U. (2015) Influence of temperature, pH, and salinity on membrane lipid composition and TEX 86 of marine planktonic thaumarchaeal isolates. *Geochimica et Cosmochimica Acta* **171**, 238–55.

Evans, T. W., Könneke, M., Lipp, J. S., et al. (2018) Lipid biosynthesis of *Nitrosopumilus maritimus* dissected by lipid specific radioisotope probing (lipid-RIP) under contrasting ammonium supply. *Geochimica et Cosmochimica Acta* **242**, 51–63.

Francis, C. A., Roberts, K. J., Beman, J. M., Santoro, A. E. and Oakley, B. B. (2005) Ubiquity and diversity of ammonia-oxidizing archaea in water columns and sediments of the ocean. *Proceedings of the National Academy of Sciences of the USA* **102**, 14683–8.

Frieling, J., Gebhardt, H., Huber, M., et al. (2017) Extreme warmth and heat-stressed plankton in the tropics during the Paleocene-Eocene Thermal Maximum. *Science Advances* **3**, e1600891.

Herfort, L., Schouten, S., Boon, J. P. and Sinninghe Damsté, J. S. (2006) Application of the TEX 86 temperature proxy to the southern North Sea. *Organic Geochemistry* **37**, 1715–26.

Hoefs, M., Schouten, S., De Leeuw, J., King, L. L., Wakeham, S. G. and Sinninghe Damsté, J. S. (1997) Ether lipids of planktonic archaea in the marine water column. *Applied and Environmental Microbiology* **63**, 3090–5.

Hollis, C. J., Dunkley Jones, T., Anagnostou, E., et al. (2019) The DeepMIP contribution to PMIP4: Methodologies for selection, compilation and analysis of latest Paleocene and early Eocene climate proxy data, incorporating version 0.1 of the DeepMIP database. *Geoscientific Model Development* **12**, 3149–206.

Hollis, C. J., Taylor, K. W. R., Handley, L., et al.(2012) Early Paleogene temperature history of the Southwest Pacific Ocean: Reconciling proxies and models. *Earth and Planetary Science Letters* **349**–350, 53–66.

Hopmans, E. C., Schouten, S., Pancost, R. D., van der Meer, M. T. and Sinninghe Damsté, J. S. (2000) Analysis of intact tetraether lipids in archaeal cell material and sediments by high performance liquid chromatography/ atmospheric pressure chemical ionization mass spectrometry. *Rapid Communications in Mass Spectrometry* **14**, 585–9.

Hopmans, E. C., Schouten, S. and Sinninghe Damsté, J. S. (2016) The effect of improved chromatography on GDGT-based palaeoproxies. *Organic Geochemistry* **93**, 1–6.

Hopmans, E. C., Weijers, J. W., Schefuß, E., Herfort, L., Sinninghe Damsté, J. S. and Schouten, S. (2004) A novel proxy for terrestrial organic matter in sediments based on branched and isoprenoid tetraether lipids. *Earth and Planetary Science Letters* **224**, 107–16.

Horak, R.E., Qin, W., Schauer, A.J., (2013) Ammonia oxidation kinetics and temperature sensitivity of a natural marine community dominated by Archaea. *The ISME Journal*, **7**, 2023.

Huguet, C., Cartes, J. E., Sinninghe Damsté, J. S. and Schouten, S. (2006) Marine crenarchaeotal membrane lipids in decapods: Implications for the TEX$_{86}$ paleothermometer. *Geochemistry, Geophysics, Geosystems* **7**. https://doi.org/10.1029/2006GC001305

Huguet, C., Kim, J.-H., de Lange, G. J., Sinninghe Damsté, J. S. and Schouten, S. (2009) Effects of long term oxic degradation on the U37 K′, TEX86 and BIT organic proxies. *Organic Geochemistry* **40**, 1188–94.

Huguet, C., Martrat, B., Grimalt, J. O., Sinninghe Damsté, J. S. and Schouten, S. (2011) Coherent millennial-scale patterns in U37 k′ and TEX86 H temperature records during the penultimate interglacial-to-glacial cycle in the western Mediterranean. *Paleoceanography* **26**. DOI: 10.1029/2010PA002048.

Hurley, S. J., Elling, F. J., Könneke, M., et al. (2016) Influence of ammonia oxidation rate on thaumarchaeal lipid composition and the TEX86 temperature proxy. *Proceedings of the National Academy of Sciences of the USA* **113**, 7762–7.

Inglis, G. N., Farnsworth, A., Lunt, D., et al. (2015) Descent toward the Icehouse: Eocene sea surface cooling inferred from GDGT distributions. *Paleoceanography* **30**, 1000–20.

Karner, M. B., DeLong, E. F. and Karl, D. M. (2001) Archaeal dominance in the mesopelagic zone of the Pacific Ocean. *Nature* **409**, 507–10.

Kim, J. G., Park, S. J., Sinninghe Damsté, J. S., et al. (2016) Hydrogen peroxide detoxification is a key mechanism for growth of ammonia-oxidizing archaea. *Proceedings of the National Academy of Sciences of the USA*, **113**, 7888–93.

Kim, J.-H., Schouten, S., Hopmans, E. C., Donner, B. and Sinninghe Damsté, J. S. (2008) Global sediment core-top calibration of the TEX86 paleothermometer in the ocean. *Geochimica et Cosmochimica Acta* **72**, 1154–73.

Kim, J.-H., Schouten, S., Rodrigo-Gámiz, M., et al. (2015) Influence of deep-water derived isoprenoid tetraether lipids on the paleothermometer in the Mediterranean Sea. *Geochimica et Cosmochimica Acta* **150**, 125–41.

Kim, J.-H., Van der Meer, J., Schouten, S., et al. (2010) New indices and calibrations derived from the distribution of crenarchaeal isoprenoid tetraether lipids: Implications for past sea surface temperature reconstructions. *Geochimica et Cosmochimica Acta* **74**, 4639–54.

Kitzinger, K., Padilla, C. C., Marchant, H. K., et al. (2019) Cyanate and urea are substrates for nitrification by Thaumarchaeota in the marine environment. *Nature Microbiology* **4**, 234.

Könneke, M., Bernhard, A. E., José, R., Walker, C. B., Waterbury, J. B. and Stahl, D. A. (2005) Isolation of an autotrophic ammonia-oxidizing marine archaeon. *Nature* **437**, 543–6.

Lengger, S. K., Hopmans, E.C., Sinninghe Damsté, J. S. and Schouten, S. (2014a) Fossilization and degradation of archaeal intact polar tetraether lipids in deeply buried marine sediments (Peru Margin). *Geobiology* **12**, 212–20.

Lengger, S. K., Hopmans, E. C., Sinninghe Damsté, J. S. and Schouten, S. J. G. (2014b) Fossilization and degradation of archaeal intact polar tetraether lipids in deeply buried marine sediments (Peru Margin). *Geobiology* **12**, 212–20.

Lengger, S. K., Sutton, P. A., Rowland, S. J., et al. (2018) Archaeal and bacterial glycerol dialkyl glycerol tetraether (GDGT) lipids in environmental samples by high temperature-gas chromatography with flame ionisation and time-of-flight mass spectrometry detection. *Organic Geochemistry* **121**, 10–21.

Lincoln, S. A., Wai, B., Eppley, J. M., Church, M. J., Summons, R. E. and DeLong, E. F. (2014) Planktonic Euryarchaeota are a significant source of archaeal tetraether lipids in the ocean. *Proceedings of the National Academy of Sciences of the USA* **111**, 9858–63.

Liu, X. L., Lipp, J. S., Birgel, D., Summons, R. E. and Hinrichs, K. U. (2018) Predominance of parallel glycerol arrangement in archaeal tetraethers from marine sediments: Structural features revealed from degradation products. *Organic Geochemistry* **115**, 12–23.

Liu, Z., Pagani, M., Zinniker, D., et al. (2009) Global cooling during the Eocene-Oligocene climate transition. *Science* **323**, 1187–90.

O'Brien, C. L., Robinson, S. A., Pancost, R. D., et al. (2017) Cretaceous sea-surface temperature evolution: Constraints from TEX 86 and planktonic foraminiferal oxygen isotopes. *Earth Science Reviews* **172**, 224–47.

Pearson, A., Hurley, S. J., Walter, S. R. S., Kusch, S., Lichtin, S. and Zhang, Y. G. (2016) Stable carbon isotope ratios of intact GDGTs indicate heterogeneous sources to marine sediments. *Geochimica et Cosmochimica Acta* **181**, 18–35.

Pearson, A., McNichol, A. P., Benitez-Nelson, B. C., Hayes, J. M. and Eglinton, T. I. (2001) Origins of lipid biomarkers in Santa Monica Basin surface sediment: A case study using compound-specific Δ14 C analysis. *Geochimica et Cosmochimica Acta* **65**, 3123–37.

Pearson, A., Pi, Y., Zhao, W., et al. (2008) Factors controlling the distribution of archaeal tetraethers in terrestrial hot springs. *Applied and Environmental Microbiology* **74**, 3523–32.

Pearson, P. N., van Dongen, B. E., Nicholas, C. J., et al. (2007) Stable warm tropical climate through the Eocene Epoch. *Geology* **35**, 211–14.

Pitcher, A., Rychlik, N., Hopmans, E. C., et al. (2010) Crenarchaeol dominates the membrane lipids of Candidatus Nitrososphaera gargensis, a thermophilic Group I. 1b Archaeon. *The ISME Journal* **4**, 542.

Polik, C. A., Elling, F. J. and Pearson, A. (2018) Impacts of paleoecology on the TEX86 sea surface temperature proxy in the Pliocene-Pleistocene Mediterranean Sea. *Paleoceanography and Paleoclimatology* **33**, 1472–89.

Qin, W., Carlson, L. T., Armbrust, E. V., et al. (2015) Confounding effects of oxygen and temperature on the TEX86 signature of marine Thaumarchaeota. *Proceedings of the National Academy of Sciences of the USA* **112**, 10979–84.

Richey, J. N. and Tierney, J. E. (2016) GDGT and alkenone flux in the northern Gulf of Mexico: Implications for the TEX86 and UK'37 paleothermometers. *Paleoceanography* **31**, 1547–61.

Robinson, S. A., Ruhl, M., Astley, D. L., (2017) Early Jurassic North Atlantic sea-surface temperatures from TEX86 palaeothermometry. *Sedimentology* **64**, 215–30.

Sagoo, N., Valdes, P., Flecker, R. and Gregoire, L. J. (2013) The Early Eocene equable climate problem: Can perturbations of climate model parameters identify possible solutions? *Philosophical Transactions of the Royal Society A: Mathematical, Physical and Engineering Sciences* **371**.

Schouten, S., Forster, A., Panoto, F. E. and Sinninghe Damsté, J. S. (2007) Towards calibration of the TEX 86 palaeothermometer for tropical sea surface temperatures in ancient greenhouse worlds. *Organic Geochemistry* **38**, 1537–46.

Schouten, S., Hopmans, E. C., Baas, M., et al. (2008) Intact membrane lipids of "*Candidatus Nitrosopumilus maritimus*," a cultivated representative of the cosmopolitan mesophilic Group I Crenarchaeota. *Applied and Environmental Microbiology* **74**, 2433–40.

Schouten, S., Hopmans, E. C., Forster, A., van Breugel, Y., Kuypers, M. M. and Sinninghe Damsté, J. S. (2003) Extremely high sea-surface temperatures at low latitudes during the middle Cretaceous as revealed by archaeal membrane lipids. *Geology* **31**, 1069–72.

Schouten, S., Hopmans, E. C., Rosell-Melé, A., et al. (2013) An interlaboratory study of TEX86 and BIT analysis of sediments, extracts, and standard mixtures. *Geochemistry, Geophysics, Geosystems* **14**, 5263–85.

Schouten, S., Hopmans, E. C., Schefuß, E. and Sinninghe Damsté, J. S. (2002) Distributional variations in marine crenarchaeotal membrane lipids: A new tool for reconstructing ancient sea water temperatures? *Earth and Planetary Science Letters* **204**, 265–74.

Schouten, S., Hopmans, E. C. and Sinninghe Damsté, J. S. (2004) The effect of maturity and depositional redox conditions on archaeal tetraether lipid palaeothermometry. *Organic Geochemistry* **35**, 567–71.

Schouten, S., Pitcher, A., Hopmans, E. C., Villanueva, L., van Bleijswijk, J. and Sinninghe Damsté, J. S. (2012) Intact polar and core glycerol dibiphytanyl glycerol tetraether lipids in the Arabian Sea oxygen minimum zone: I. Selective preservation and degradation in the water column and consequences for the TEX86. *Geochimica et Cosmochimica Acta* **98**, 228–43.

Schouten, S., Villanueva, L., Hopmans, E.C., van der Meer, M.T. and Damsté, J.S.S., 2014. Are Marine Group II Euryarchaeota significant contributors to tetraether lipids in the ocean?. Proceedings of the National Academy of Sciences, 111(41), pp.E4285–E4285, https://doi.org/10.1073/pnas.1416176111

Shah, S. R., Mollenhauer, G., Ohkouchi, N., Eglinton, T. I. and Pearson, A. (2008) Origins of archaeal tetraether lipids in sediments: Insights from radiocarbon analysis. *Geochimica et Cosmochimica Acta* **72**, 4577–94.

Sinninghe Damsté, J. S., Rijpstra, W. I. C., Hopmans, E. C., den Uijl, M. J., Weijers, J. W. and Schouten, S. (2018) The enigmatic structure of the crenarchaeol isomer. *Organic Geochemistry*, **124**, 22–8.

Sinninghe Damsté, J. S., Schouten, S., Hopmans, E. C., Van Duin, A. C. and Geenevasen, J. A. (2002) Crenarchaeol the characteristic core glycerol dibiphytanyl glycerol tetraether membrane lipid of cosmopolitan pelagic crenarchaeota. *Journal of Lipid Research* **43**, 1641–51.

Taylor, K. W., Huber, M., Hollis, C. J., Hernandez-Sanchez, M. T. and Pancost, R. D. (2013) Re-evaluating modern and Palaeogene GDGT distributions: Implications for SST reconstructions. *Global and Planetary Change* **108**, 158–74.

Tierney, J.E. and Tingley, M. P. (2014) A Bayesian, spatially-varying calibration model for the TEX$_{86}$ proxy. *Geochimica et Cosmochimica Acta* **127**, 83–106.

Tierney, J. E. and Tingley, M. P. (2015) A TEX86 surface sediment database and extended Bayesian calibration. *Scientific Data* **2**.

Turich, C., Freeman, K. H., Bruns, M. A., Conte, M., Jones, A. D. and Wakeham, S. G. (2007) Lipids of marine Archaea: Patterns and provenance in the water-column and sediments. *Geochimica et Cosmochimica Acta* **71**, 3272–91.

Uda, I., Sugai, A., Itoh, Y. H. and Itoh, T. (2001) Variation in molecular species of polar lipids from *Thermoplasma acidophilum* depends on growth temperature. *Lipids* **36**, 103–5.

Villanueva, L., Schouten, S. and Sinninghe Damsté, J. S. (2015) Depth-related distribution of a key gene of the tetraether lipid biosynthetic pathway in marine Thaumarchaeota. *Environmental Microbiology* **17**, 3527–39.

Weijers, J. W., Schouten, S., Spaargaren, O. C. and Sinninghe Damsté, J. S. (2006) Occurrence and distribution of tetraether membrane lipids in soils: Implications for the use of the TEX86 proxy and the BIT index. *Organic Geochemistry* **37**, 1680–93.

Wörmer, L., Elvert, M., Fuchser, J., et al. (2014) Ultra-high-resolution paleoenvironmental records via direct laser-based analysis of lipid biomarkers in sediment core samples. *Proceedings of the National Academy of Sciences of the USA* **111**, 15669–74.

Wuchter, C., Schouten, S., Boschker, H. T. and Sinninghe Damsté, J. S. (2003) Bicarbonate uptake by marine Crenarchaeota. *FEMS Microbiology Letters* **219**, 203–7.

Wuchter, C., Schouten, S., Coolen, M. J. L. and Sinninghe Damsté, J. S. (2004) Temperature-dependent variation in the distribution of tetraether membrane lipids of marine Crenarchaeota: Implications for TEX86 paleothermometry. *Paleoceanography* **19**, PA4028.

Wuchter, C., Schouten, S., Wakeham, S. G. and Sinninghe Damsté, J. S. (2005) Temporal and spatial variation in tetraether membrane lipids of marine Crenarchaeota in particulate organic matter: Implications for TEX86 paleothermometry. *Paleoceanography* **20**.

Wuchter, C., Schouten, S., Wakeham, S. G. and Sinninghe Damsté, J. S. (2006) Archaeal tetraether membrane lipid fluxes in the northeastern Pacific and the Arabian Sea: implications for TEX86 paleothermometry. *Paleoceanography* **21**.

Yamamoto, M., Shimamoto, A., Fukuhara, T., Tanaka, Y. and Ishizaka, J. (2012) Glycerol dialkyl glycerol tetraethers and TEX86 index in sinking particles in the western North Pacific. *Organic Geochemistry* **53**, 52–62.

Zachos, J. C., Schouten, S., Bohaty, S., et al.(2006) Extreme warming of mid-latitude coastal ocean during the Paleocene-Eocene Thermal Maximum: Inferences from TEX86 and isotope data. *Geology* **34**, 737–40.

Zeng, Z., Liu, X. L., Farley, K. R., et al. (2019) GDGT cyclization proteins identify the dominant archaeal sources of tetraether lipids in the ocean. *Proceedings of the National Academy of Sciences of the USA* **116**, 22505–11.

Zhang, Y. G. and Liu, X. (2018) Export depth of the TEX86 signal. *Paleoceanography* **33**, 666–71.

Zhang, Y. G., Pagani, M. and Wang, Z. (2016) Ring Index: A new strategy to evaluate the integrity of TEX86 paleothermometry. *Paleoceanography* **31**, 220–32.

Zhang, Y. G., Zhang, C. L., Liu, X.-L., Li, L., Hinrichs, K.-U. and Noakes, J. E. (2011) Methane Index: a tetraether archaeal lipid biomarker indicator for detecting the instability of marine gas hydrates. *Earth and Planetary Science Letters* **307**, 525–34.

Zhou, A., Weber, Y., Chui, B. K., et al. (2019) Energy flux controls tetraether lipid cyclization in *Sulfolobus acidocaldarius*. *Environmental Microbiology* **22**, 343–53.

Cambridge Elements $\equiv$

Elements in Geochemical Tracers in Earth System Science

Timothy Lyons

University of California

Timothy Lyons is a Distinguished Professor of Biogeochemistry in the Department of Earth Sciences at the University of California, Riverside. He is an expert in the use of geochemical tracers for applications in astrobiology, geobiology and Earth history. Professor Lyons leads the 'Alternative Earths' team of the NASA Astrobiology Institute and the Alternative Earths Astrobiology Center at UC Riverside.

Alexandra Turchyn

University of Cambridge

Alexandra Turchyn is a University Reader in Biogeochemistry in the Department of Earth Sciences at the University of Cambridge. Her primary research interests are in isotope geochemistry and the application of geochemistry to interrogate modern and past environments.

Chris Reinhard

Georgia Institute of Technology

Chris Reinhard is an Assistant Professor in the Department of Earth and Atmospheric Sciences at the Georgia Institute of Technology. His research focuses on biogeochemistry and paleoclimatology, and he is an Institutional PI on the 'Alternative Earths' team of the NASA Astrobiology Institute.

About the series

This innovative series provides authoritative, concise overviews of the many novel isotope and elemental systems that can be used as 'proxies' or 'geochemical tracers' to reconstruct past environments over thousands to millions to billions of years—from the evolving chemistry of the atmosphere and oceans to their cause-and-effect relationships with life.

Covering a wide variety of geochemical tracers, the series reviews each method in terms of the geochemical underpinnings, the promises and pitfalls, and the 'state-of-the-art' and future prospects, providing a dynamic reference resource for graduate students, researchers and scientists in geochemistry, astrobiology, paleontology, paleoceanography and paleoclimatology.

The short, timely, broadly accessible papers provide much-needed primers for a wide audience—highlighting the cutting-edge of both new and established proxies as applied to diverse questions about Earth system evolution over wide-ranging time scales.

Cambridge Elements ☰

Elements in Geochemical Tracers in Earth System Science